The Volcano Book for Children, Mums, Dads and Teachers

Rick Lomas

The Volcano Book for Children, Mums, Dads and Teachers

By Rick Lomas
Edited by Susan Lomas

Version CS1.0 UK and Europe version

ISBN-13: 978-1492390633

RickLomas.com Publishing

September 2013

RickLomas.com

Dedication

This book is dedicated to my children, Madeline and Rosalie and their Mum, Susan.

Thanks also to my Mum and Dad for introducing me to Lanzarote and its wonderful volcanic landscape.

CONTENTS

Contents

Did you know?

This book is also available as a Amazon Kindle book (ASIN: B00F2JGJ36).

If you have already bought this paperback copy of the book from Amazon you will automatically be entitled to a reduced price (and sometimes FREE) Kindle version of this book.

Rick and Madeline Lomas on holiday in Lanzarote
(Volcano Island) 30th October 2005

I wrote this book because my eldest daughter Madeline
kept asking me questions about volcanoes. When she
was about 7 years old she was quite shocked when she
realised we had taken her on holiday to Lanzarote in
2005, when she was just one and half years old.
Lanzarote is a volcanic island that had some major
eruptions between 1730 and 1736 which made thirty-
two new volcanoes. Lava covered a quarter of
Lanzarote's surface, including good farmland and eleven
villages. Madeline always liked to call Lanzarote,
'Volcano Island'. Even when she was quite little she

would want to know how dangerous volcanoes were and where to go and see some. I guess when her younger sister Rosalie gets a little bit bigger she will be asking me similar questions too!

So this book is for children and for their Mums and Dads. Teachers will find this useful too. It is not too complicated, but at the same time I haven't skipped over any important points. It is almost impossible to talk about volcanoes without knowing what a pyroclastic flow or a lahar is, so that is all explained in here.

I was tempted to make it a nice fun book to read in a light hearted way, but the truth is that volcanoes have killed thousands and thousands of people since the earth began. The Lake Toba eruption nearly wiped humans off the planet completely! Although the danger of death by volcano is fairly low these days, it is good to know what has happened in the past and what could happen in the future. I suppose the unknown is what makes volcanoes so exciting.

I do hope you have fun reading this and learn something at the same time.

Rick Lomas, September 2013

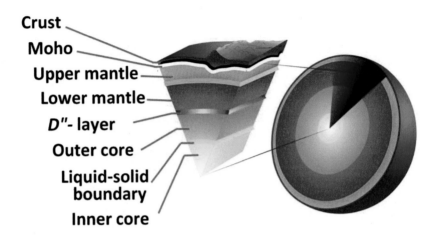

Crust
Moho
Upper mantle
Lower mantle
D"- layer
Outer core
Liquid-solid boundary
Inner core

The Earth has layers, quite a few actually!

We live on the planet Earth and it seems pretty solid doesn't it? Well you might be surprised to find out that the Earth isn't as solid as you might think. The solid part of the earth that we live on is called the Earth's crust.

The Earth's crust varies in thickness depending on where you are. If you are in England the Earth's crust is about 30 kilometres (km) thick, if you are in the middle of North America, say somewhere like Kansas, then it's a little bit thicker, about 45 km, but if you are in Iceland it's only about 10 km thick.

The part of the earth we are most concerned with here is the crust and the mantle layers. This is where volcanoes are born.

MAGMA CHAMBERS

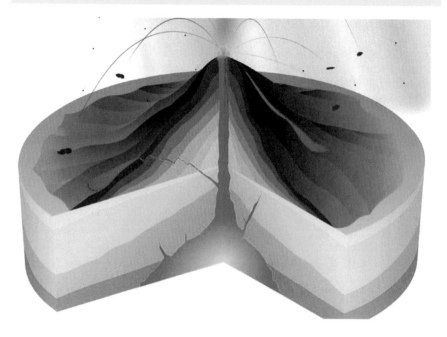

A magma chamber is like a big underground pool of molten rock

Below the crust of the Earth is a layer called the Earth's mantle. The mantle is actually pretty solid but it is very hot and some of it is a gloopy goo called magma. The magma tends to form in big underground pools, which are called magma chambers. Magma is actually a mixture of molten rock, gases and some other bits and pieces too.

WHAT ARE TECTONIC PLATES?

The Earth's crust is made of large pieces called tectonic plates

"The Earth's crust gets interesting when you find out that it is not just one big crust like the shell of an egg. In fact the Earth's crust is made out of some very big pieces and lots of smaller ones that fit together, a bit like a jigsaw. These are called tectonic plates."

There are seven big ones called primary tectonic plates, eight smaller ones called secondary tectonic plates and then lots of little ones called tertiary tectonic plates. These tectonic plates are always moving against each other, usually just a tiny amount, but sometimes they can move quite a lot. This is what causes terrible things like earthquakes and tsunamis to happen.

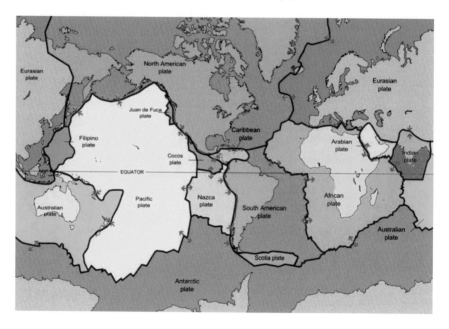

This map shows fifteen of the Earth's largest tectonic plates. Where do you live?

Use the map above to see if you live near the boundary of two tectonic plates. If you do there may be a volcano near you.

OH NO! THE EARTH'S CRUST IS BROKEN!

The 1982 eruption of Galungung in western Java, Indonesia

A volcano is made when a hole or a crack happens in the Earth's crust. As you can imagine this tends to happen more often where two tectonic plates meet each other. But volcanoes don't always happen where tectonic plates meet, they can also happen in the middle of a tectonic plate. This usually happens where the Earth's crust is thin - these are called volcano hotspots.

A typical hotspot is to be found in Hawaii, where the liquid magma is only 3 km below the Earth's surface. Another example is the Canary Islands which include Lanzarote, Tenerife, Fuerteventura and Gran Canaria.

The Canary Islands are located on the African Plate, but not at the edge. This area is known as the 'Canary Hotspot'.

When the magma finds its way through the Earth's crust it spurts out and makes a volcano. When the magma comes out of a volcano we don't call it magma any more, we call it lava. When a volcano has lava coming out of it, we say this is a volcanic eruption.

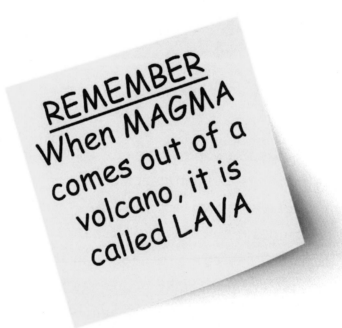

REMEMBER
When MAGMA comes out of a volcano, it is called LAVA

There are different kinds of lava; some is thin and runs like a river and some is really thick and doesn't move much at all. The temperature of magma is usually between 700 C and 1300 C, so even when it comes out of a volcano and turns into lava it is still very hot. When lava cools down it turns into rock, which is known as volcanic rock.

WHAT IS A CALDERA?

La Cumbre volcano on Fernandina Island, one of the Galapagos Islands. A perfect caldera!

One feature of many volcanoes is that they have a crater at the top, this is called a caldera. A caldera is formed when much of the magma has left the magma chamber and the rock above it collapses and causes a crater to form.

Quite often a lake may form in the caldera; these are known as caldera lakes. We will hear more about calderas when we talk about supervolcanoes.

Crater Lake is a caldera lake in Oregan, USA.

ACTIVE, DORMANT OR EXTINCT?

Generally when we talk about volcanoes we talk about them being active, dormant or extinct. These definitions do vary a little depending on who or what you are talking about. They mean different things to different people and to different volcanoes depending on who you are talking to.

We are going to use the definitions that are most used today. So let's look at what that really means, starting with the worst:

ACTIVE VOLCANOES

An active volcano, Mount Etna erupting in 2011

Yes these are the bad boys, but what that means is that they have erupted at some point since the last Ice Age, so sometime in the last 10,000 years!

DORMANT VOLCANOES

A dormant volcano, Putauaki in New Zealand

These are volcanoes that haven't erupted for over 10,000 years, but there is a possibility of another eruption.

EXTINCT VOLCANOES

You don't need to worry about these volcanoes; nobody expects them to erupt ever again!

THE RING OF FIRE

This is also referred to as the Pacific Ring of Fire. There are about 1500 volcanoes around the world which are classified as active. Nearly 90% of these are in a ring around the Pacific Ocean.

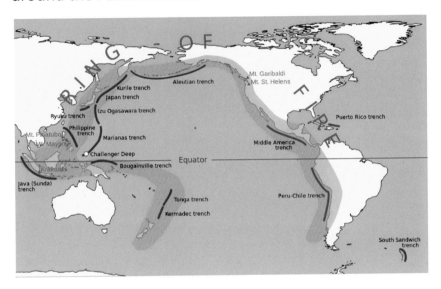

The Pacific Ring of Fire

As you can see the Ring of Fire does not really look like a ring as it doesn't join up at the bottom. However the real Earth is not flat like the map above, so New Zealand and South America are actually nearer to each other than they look.

ARE VOLCANOES DANGEROUS?

Errr yes! Volcanoes are indeed dangerous, but compared to earthquakes and tsunamis they aren't quite so bad. In the last hundred years they have caused the deaths of about 850 people a year. So how do they kill you? Here's how:

LAVA FLOW

Ten metre high fountain of lava, Hawaii, United States

As you can imagine very hot molten rock pouring out of a volcano is not good for anyone or anything nearby! However, it might surprise you to know that the lava flow is actually one of the least dangerous things about a volcano. The lava usually moves quite slowly, about the same speed as you walk, so you can just run away from it. Buildings and trees can't run away though so they might suffer a bit more. But there are worse things than the lava flow!

PYROCLASTIC FLOW

Pyroclastic flow at the Mayon volcano in The Philippines, 1984

One of the most dangerous things about an erupting volcano is the pyroclastic flow. Pyroclastic flow is like a big cloud of hot gas and rocks. It's not like a fluffy cloud that is going to float away on its own, this is a heavy cloud. The pyroclastic flow is forced down by gravity from the top of the volcano and follows the slope down the side of the volcano. The worst thing about pyroclastic flow is that it moves very quickly and it is very hot. Pyroclastic flow can travel at speeds of up to 700 km per hour and can get as hot as 1000 C.

Vesuvius, as seen from the ruins of Pompeii, which was destroyed in the eruption of AD 79

Nearly 2000 years ago in Italy a volcano called Mount Vesuvius erupted. The pyroclastic flow completely destroyed the towns of Pompeii and Herculaneum. It is believed that 16,000 people were killed by the pyroclastic flow. The total number of people killed in the disaster was 18,000.

LAHARS

A lahar from the 1982 eruption of Galunggung, West Java, Indonesia

Lahar is a nice sounding word which you would think would describe something pretty, but it is definitely not! A lahar is like a muddy river which is made out of a horrible gunge of rocky debris, pyroclastic flow and water. The lahar comes out of the volcano and flows down the side and then tends to follow rivers and valleys. Lahars destroy almost everything in their path. We will see later, when we talk about the most deadly volcanoes, how lahars have killed thousands of people.

Since 1783 lahars have been responsible for 17% of volcano related deaths. Lahars vary in speed, but some have been known to travel distances of up to 300 km at speeds of up to 100 km per hour. When a lahar is travelling fast like this they are deadly and have been known to bury cities and de-rail express trains. Nasty, very nasty.

POISONOUS GASES

When a volcano erupts there is lots of water vapour as well as carbon dioxide and sulphur dioxide, which can be very bad for you. If you are in a cloud of carbon dioxide you will probably suffocate, as there will be no air to breathe. Sulphur dioxide mixed with water vapour causes acid rain, which is about as bad as it sounds. The chemicals in acid rain can cause paint to peel and steel structures, like bridges, to corrode. Acid rain is also harmful to plants, animals and people.

VOLCANIC ASH

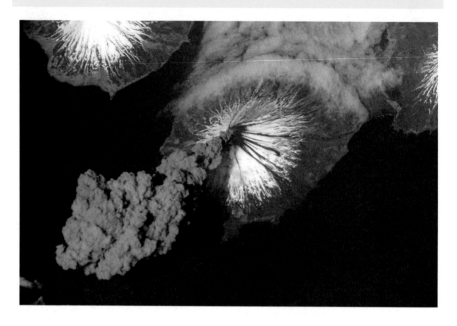

Ash plume from Mount Cleveland, Alaska

Volcanic ash is created when a volcano erupts. The ash is made out of tiny fragments of rock, minerals and volcanic glass. Volcanic glass is made as the magma from inside the volcano meets the air outside and cools down. The bits of ash are less than 2mm in size. The volcanic ash is not good for people or animals; if you breathe it in it will make you very poorly or even kill you.

Volcanic ash is also very bad for aeroplane engines. Even just one tiny piece of ash getting into an aeroplane's engine can cause the engine to stop working. So usually when a volcano erupts no aeroplanes are allowed to fly anywhere near it. But it gets worse; once the volcanic ash is in the air the wind can blow it along for hundreds or even thousands of kilometres.

In 1991 a volcano called Mount Pinatubo erupted in the Philippines and about twenty aeroplanes were damaged. These aeroplanes were more than 950 km away from the volcano. Volcanic ash from Mount Pinatubo was also found on the east coast of Africa, more than 8000 km away.

In 2010 there was a big volcano called Eyjafjallajökull that erupted in Iceland. Although compared to some other volcanic eruptions this wasn't a very big one, the volcanic ash cloud from it was so big that twenty countries did not let aeroplanes fly in their airspace. This meant that most flights in Europe had to be cancelled and this affected more than 100,000 people who wanted to travel.

Volcanic ash can cause problems if it falls on electricity power lines, especially when the ash is wet. Electricity can travel through the wet volcanic ash instead of the wires where it is supposed to go! So until everything is cleaned up the power supply has to be cut. For this reason sometimes whole towns and cities will be without electricity for quite a long time.

The ash is also very heavy so if it settles on the roof of any sort of building it can cause it to collapse. This can cause problems with homes, but it can also destroy factories and other buildings that are important, such as power stations and hospitals.

Very big volcanoes can cause so much volcanic ash and sulphur dioxide that they block the sun from the Earth and cause the planet to go cold and dark for a few years. These are called volcanic winters and can be devastating for our Earth, making it impossible to grow crops and feed ourselves.

DIFFERENT TYPES OF VOLCANO

Most of the time we think of a volcano as a big cone shaped mountain with a crater (caldera) on the top where all the lava and poisonous gases come out. This is actually just one type of volcano and there are quite a few more which I will tell you about in this chapter.

FISSURE VENTS

The Laki Fissure in Iceland, not the prettiest of volcanoes, but one of the most deadly!

Fissure vents are quite easy to explain as they are just a long crack in the Earth's crust that has some lava seeping out of it. This sounds like quite a boring volcano, but it's not. A good example is the Lakagigar Fissure in Iceland.

Lakagigar actually means 'Craters of Laki' and Laki is the name of the mountain.

Back in the year 1783 the Laki mountain itself didn't erupt, but on either side of it cracks, known as fissures, opened up. Lakagigar is part of a volcano system which includes the Grimsvötn volcano which also erupted at the same time. During 1783 the Lakagigar Fissure and the Grimsvötn volcano managed to pour out around 14 km³ of lava over an eight-month period; that's a lot of lava!

The poisonous gases coming out of the Lakagigar Fissure and the Grimsvötn volcano were sulphur dioxide and hydrofluoric acid. These gases killed half of the farm animals in Iceland, which meant that food was in short supply. A quarter of the people in Iceland died of hunger. As the sulphur dioxide went into the air it caused the average temperature to drop which went on to cause more problems. Farmers couldn't grow their crops in Europe and some believe it may even have caused similar problems as far as India. We now believe this eruption may have killed as many as six million people, which makes it the deadliest eruption in history.

SHIELD VOLCANOES

The volcano Skjaldbreiður in Iceland; a perfect example of a shield volcano

Shield volcanoes are very pretty. They are called shield volcanoes because they look like a soldier's circular shield lying on the ground, although I think they look a bit more like a cymbal from a drum kit. They are this shape because the lava that comes out of them is generally quite runny and flows for a long distance quite gently. Shield volcanoes are common in Iceland and these are also the ones you will find in Hawaii.

LAVA DOMES

Lava domes in the crater of Mount St. Helens, Washington State, USA

Lava domes are formed when the lava is very thick and the eruption is quite slow. However lava domes can be quite dangerous as they may produce explosive eruptions. Sometimes you can see lava domes inside the craters of older volcanoes that have already erupted. One example of this is Mount St Helens in the United States. The good thing about lava domes is the fact that the lava doesn't normally travel very far.

CRYPTODOMES

The bulging cryptodome on the side of Mount St Helens on April 27th, 1980

Cryptodomes are funny looking volcanoes where you actually see a bulge on the Earth before the lava breaks through. Sometimes they appear on the side of a mountain and get bigger and bigger and then explode. This is what happened at Mount St Helens, a famous volcano in the USA's Washington State, during 1980.

The man in the photograph with the parasol is measuring how much that bulge is moving each day. The mountain may look solid, but Mount St Helens changes shape, a little bit, every day.

VOLCANIC CONES OR CINDER CONES

Sunset Crater near Flagstaff, Arizona, USA

Volcanic cones or cinder cones tend to be quite small, measuring from about 30 metres high up to 400 metres high. Usually they only erupt once and the eruption does not last for a very long time. Sunset Crater near Flagstaff, Arizona, in the United States, is a very beautiful example of a cinder cone.

STRATOVOLCANOES OR COMPOSITE VOLCANOES

Stromboli, in Italy, is good example of a stratovolcano. Photo taken by Patrick Nouhailler on June 20th 2013

Stratovolcanoes are tall and conical and are made in layers. As all the ash and lava comes out of the volcano it cools and hardens. This happens again and again. Each time when the lava cools down it forms a new layer on top of the one below it. These sorts of volcanoes are considered to be one of the most dangerous because they are always very steep which means that the lava and the pyroclastic flow can move much quicker down the side of the volcano. Some good examples of stratovolcanoes are Mount Vesuvius and Stromboli in Italy.

SUPERVOLCANOES

The Yellowstone Caldera is the caldera and supervolcano located in Yellowstone National Park, USA

As you might have guessed by their name supervolcanoes are big, very big. Guess what? They are very dangerous too.

A supervolcano can cause devastation across whole countries or even continents during and after eruption. The ash cloud prevents the sun's rays from reaching the Earth for a period of time causing volcanic winters. As this affects global temperatures for years after the eruption, supervolcanoes are in many ways the most dangerous type of volcano.

One of the best examples of a supervolcano is the Yellowstone Supervolcano in the Rocky Mountains in the United States. At the top of this volcano is the huge caldera which measures 55 km x 72 km.

The Volcano Book for Children, Mums, Dads and Teachers

DIFFERENT TYPES OF VOLCANIC ERUPTIONS

Not only are there different types of volcanoes but there are different types of volcanic eruptions too. A volcano can erupt in different ways during active periods. There are quite a few interesting names for these eruptions, but the ones you hear about most are: Hawaiian, Strombolian, Vulcanian, Peléan and Plinian. I have listed these in order of nastiness, starting with the not so nasty Hawaiian eruption, leading right up to the very nasty Plinian eruption.

HAWAIIAN ERUPTIONS

Hawaiian eruptions are named after the type of volcanic eruptions you get in Hawaii, but they don't just happen in Hawaii. In many ways Hawaiian eruptions are the least scary volcanic activity although that's not to say that they don't kill people because they certainly do. Hawaiian eruptions tend to happen little and often to particular volcanoes. The amount of gas they give off is quite small and the lava tends to come out quite slowly and doesn't travel too far.

STROMBOLIAN ERUPTIONS

Strombolian eruptions are named after a volcano in Sicily called Stromboli. Strombolian eruptions can last for a long time. Stromboli itself has been erupting for thousands of years. The lava from a Strombolian eruption is thicker and gloopier than the lava from a Hawaiian eruption, so it tends to travel a bit slower and not as far. Strombolian eruptions are more dangerous than Hawaiian eruptions as the thick lava can turn quite solid in places. The pressure behind the blockage can then cause explosions at fairly regular intervals, spurting out bombs of lava a few hundred metres into the air.

VULCANIAN ERUPTIONS

The word Vulcanian comes from the island of Vulcano which is about 25 km north of Sicily. Vulcanian eruptions can be pretty nasty. This sort of volcanic eruption is always characterised by a big, thick dense cloud of ash rising from the top of the volcano. They can also explode over and over again which can sound like a load of cannons being fired. These explosions are actually the pyroclastic flow being shot out by explosions of gas.

Sometimes the pyroclastic flow can even travel faster than the speed of sound creating a sonic boom, just like the sonic boom from a jet fighter plane as it reaches a supersonic speed. Vulcanian eruptions can also throw out large lumps of rock about one metre across which can then travel for hundreds of metres. Many people have died or been injured by rocks of this kind..

PELÉAN ERUPTIONS

Peléan eruptions have some similarities to Vulcanian eruptions. These types of eruptions have terrible pyroclastic flows and an avalanche of hot volcanic ash. They also tend to form lava domes. Normally these eruptions only last for a few years. A good example of a Peléan eruption is the 1980 eruption of Mount St Helens in the United States. The 1980 eruption killed fifty-seven people and thousands of animals. It spread ash over eleven American states causing billions of dollars of damage.

There is nothing nice about a Peléan eruption, but it is not the worst kind...

PLINIAN ERUPTIONS

Plinian eruptions are as bad as you can get. They are also sometimes called Vesuvian eruptions. These eruptions take their names from a letter that was written by Pliny the Younger whose uncle, Pliny the Elder, was killed by the eruption of Vesuvius in AD 79. This was during the time of the Roman Empire in Italy. It is believed that Pliny the Elder died from either suffocation or poisoning in a cloud of volcanic gas.

Plinian eruptions make tall columns of gas and rocks shoot out from the top of the volcano into the atmosphere. The blasts of gas are very powerful and happen continuously. Some short eruptions only last for about a day, but others have been known to last for several days or even months. Because there is so much magma coming out of the volcano the magma chamber begins to get empty and then the top of the volcano can collapse forming a caldera.

THE NICE THINGS ABOUT VOLCANOES

So far we have just talked about how bad volcanoes are, so now let's give them a chance! Volcanoes can cause damage and destruction but over a long period of time they also have benefited people. Here are some nice things about volcanoes:

VOLCANOES CREATED 80% OF THE EARTH'S SURFACE AND MOST OF THE AIR WE BREATHE TODAY

Yes, indeed this is true. The sea floor and some mountains were formed by volcanic eruptions which happened over and over again. The Earth's crust is mostly the product of millions of volcanoes that were once active. Large quantities of magma did not erupt but instead cooled below the surface. Even though there are a lot of horrible gases that come from a volcano over time these gases have broken down and formed the Earth's atmosphere.

VOLCANOES CREATE NEW FERTILE LAND THAT CAN BE USED FOR FARMING

Over thousands to millions of years, the breakdown and chemical changing of volcanic rocks have formed some of the most fertile soils on Earth. Take for example a tropical, rainy place such as the north east side of the island of Hawaii where the formation of fertile soil and the growth of lush plants after an eruption can happen within just a few hundred years. Some of the earliest civilizations around Greece and Italy decided to live on the rich, fertile volcanic soils in the Mediterranean and Aegean regions. Some of the best rice growing regions of Indonesia are right next to active volcanoes. Also, many of the best farming areas in the western United States have fertile soils which have come from volcanic rocks.

GEOTHERMAL ENERGY

Some countries use the heat below the Earth's surface to create electricity and heating systems using geothermal energy. In Iceland, geothermal heat warms more than 70 % of the homes in the country.

'The Geysers' geothermal power plant in the Mayacamas mountains, Sonoma County, California, USA

116 km north of San Francisco, in the Mayacamas mountains near Santa Rosa, is a complex of twenty-two geothermal power plants, which use the steam from more than three hundred and fifty underground wells. The complex is called 'The Geysers' and this geothermal field in northern California can produce enough electricity to power San Francisco.

VOLCANOES ARE A SOURCE OF MINERALS

Most of the minerals that are mined on our planet such as copper, gold, silver, lead, and zinc come from the magma which is found deep inside extinct volcanoes. Rising magma does not always get to the surface to erupt; sometimes it cools slowly and hardens beneath the volcano to form crystalline rocks. With the right temperature and pressure precious minerals can form.

Here are some volcanoes that have been important in the history of the Earth and are quite easy to visit if you are in that part of the world.

STROMBOLI, ITALY

Stomboli volcano, photo by Patrick Nouhailler, June 20th 2013

Stromboli is sometimes known as the 'Lighthouse of the Mediterranean' because of its spectacular explosions that you can see at night. The Stromboli eruptions now happen very often and the word Stromboli is used to mean 'small explosion'. If you want you can get very near to the active crater!

Stromboli is on one of the Aeolian Islands near Sicily. You can only get to Stromboli by boat because there is nowhere to put an airport. There are hydrofoil and ferry services which connect Stromboli's island to other places in the Aeolian Islands and also to Sicily and the mainland of Italy.

KILAUEA, HAWAII

Red hot lava oozes into the ocean at the Kilauea volcano, Hawaii on April 4th 2008. Photo by Bob Webster

Hawaii is a long way away from most places, even if you live in California it is still a five hour flight from Los Angeles. But if you ever find yourself in Hawaii, you are in for a volcanic treat!

Kilauea on Hawaii's largest island is now said to be the world's most active volcano. If you want to be sure of seeing some volcano action, this is the place to go. You can even take a boat ride to see the red hot lava oozing into the ocean where it instantly turns the sea into steam. That really is something you don't see every day!

MOUNT ETNA, SICILY, ITALY

Burnt out trees in a lava field, Mount Etna. Photo by 'gnuckx', May 1st 2009

If you can't get to Hawaii why not visit Europe's most active volcano instead? Mount Etna, on the island of Sicily at the foot of Italy, is always doing something. To see it close up you can take a 4x4 trip to explore Etna's lava fields with caves, tunnels and fissures. If it's a fiery volcano you want, Mount Etna is perfect.

TEIDE, TENERIFE, SPAIN

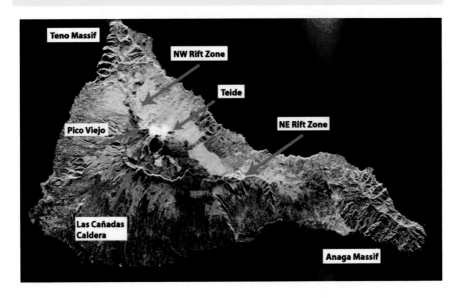

Space radar image of Teide, taken in 1994 by NASA

If you prefer your volcanoes a little less fiery than Etna, maybe a visit to Mount Teide in Tenerife is the place to go, especially as it hasn't erupted since 1909. Tenerife is part of the Canary Islands, which are just off the coast of North Africa but actually belong to Spain. The peak of Teide is 3718 metres which means it is the highest point in Spain. You can take the Mount Teide cable car to the summit to enjoy the view, but don't forget to apply for a permit first.

Tenerife is quite a popular destination for holidaymakers from the UK and Europe. So if you live in that part of the World it is quite likely that you could find yourself in

Tenerife. If you go to take a look at Mount Teide don't do what I did and go dressed for the beach. At that altitude it gets a bit chilly, even in the summer!

MOUNT RUAPEHU, NEW ZEALAND

Mount Ruapehu and Chateau Tongariro.
Photo taken on September 14th 2009 by Sid Mosdell

This volcano is in part of the Pacific Ring of Fire. You will find a good number of active volcanoes in New Zealand. Mount Ruapehu is the country's highest volcano at 2,797m and in the winter you can go skiing there. It is also one of the world's most active volcanoes, having last erupted on the 25th of September 2007.

Mount Ruapehu has three major peaks: the biggest is Tahurangi at 2,797 m, then Te Heuheu at 2,755 m and lastly the one with the longest name, Paretetaitonga at 2,751 m. These peaks overlook a deep, active crater. Between major eruptions a caldera lake forms and this can cause some problems.

In 1945 an eruption emptied Mount Ruapehu's caldera lake and then blocked its outlet by forming a dam with volcanic debris. The crater slowly filled up with water from melting snow, until on the 24th of December 1953 the dam collapsed causing a lahar in the Whangaehu River.

The lahar caused the Tangiwai railway bridge across the Whangaehu River to collapse, just before a train was about to cross it. The train driver was warned about the collapsed bridge, but was unable to stop the train in time and six of the carriages fell into the river. This was the Tangiwai disaster, which killed 151 people.

MOUNT FUJI, JAPAN

Mount Fuji, Japan.
Photo taken on May 3rd 2010 by Ivan Walsh

Mount Fuji is on the island of Honshu and is the highest mountain in Japan at 3,776m. It is a strato-volcano and is still classed as active, although it hasn't erupted since 1708, so it is probably pretty safe.

So would you fancy hiking up Mount Fuji? It is a popular volcano to hike up, not only for Japanese people but also foreign tourists, who make up more than a third of its visitors.

If you get the opportunity to hike up Mount Fuji be aware that at some points the ground is steep and rocky. There are lots of signs which warn you about possible dangers like sudden gusts of wind and falling rocks. The real challenge of climbing Mount Fuji is the fact that it is a very strenuous ascent and breathing gets more difficult as you get higher and the air gets thinner.

COTOPAXI, ECUADOR

Cotopaxi volcano, Ecuador, South America.
Photo taken on September 26th 2009
by Ivar Abrahamsen

In the Andes Mountains in Ecuador there is an area called the 'Avenue of the Volcanoes' which includes the Cotopaxi volcano. At almost 6000 metres above sea level, Cotopaxi is the world's highest active volcano.

Since 1738, Cotopaxi has erupted more than fifty times. Around the volcano there are lots of valleys that have been formed by lahars. It is possible that Cotopaxi may erupt again and this could be devastating for the local population. A huge eruption at Cotopaxi might cause a dangerous flow of ice from its glacier. If the explosion was big it would easily destroy homes and farming land in the valley below which has a population of about 2,000,000 people. This happened to another city nearby called Latacunga which was destroyed in the 18th century by volcanic activity.

SIERRA NEGRA, GALAPAGOS ISLANDS

The Sierra Negra volcano.
Photo taken by Michael R Perry, December 23rd 2009

This is one of the largest volcanoes in the Galapagos Archipelago. The Sierra Negra is a large shield volcano with an altitude of 1124m. As this is one of the most active volcanic areas in the world, the Galapagos Islands are a great place for volcano lovers to visit and hopefully witness some volcanic activity.

The last eruption started on October 22nd 2005 and finished on October 30th 2005. Even though scientists are always keeping an eye on Sierra Negra there was no warning of this eruption.

One of the reasons I would like to visit this volcano is that it is the natural habitat for the gorgeous Sierra Negra giant tortoise, which lives on the slopes of the volcano. Although there is a restoration program in place to preserve these tortoises they are an endangered species. Over the years they have been hunted and been preyed upon by other animals. We need to look after these giant tortoises; an eruption of Sierra Negra may not be their biggest threat.

YELLOWSTONE SUPERVOLCANO, WYOMING, USA

A geyser, where water is turned instantly to steam, at Yellowstone, October 13th 2000

Although the Yellowstone supervolcano's last eruption happened nearly 640,000 years ago, the area still has a lot of volcanic activity. Visitors come to Yellowstone Park to see the geysers shooting out their steam. If Yellowstone did erupt again, scientists say it could cause massive devastation all over the United States and Canada. However it is reckoned that this won't happen for about another 90,000 years, so we are OK for the moment.

There are always a lot of earthquakes in the Yellowstone area (between 1000 and 2000 earthquakes a year), although most are very small.

MONTSERRAT, WEST INDIES

Some of the buildings damaged by the volcano near Plymouth, Montserrat

In 1995 the Soufrière Hills volcano erupted, burying and burning the former capital of Plymouth to such an extent that they had to relocate the government permanently.

Ash continues to fall to this day making much of the area off limits, but some parts of the exclusion zone are now open for guided tours allowing visitors to see the half-buried houses.

The Soufrière Hills volcano is best viewed from Jack Boy Hill from where you can watch its angry red glow after dark.

MOUNT PELÉE, MARTINIQUE

Main Street, Morne Rouge, Martinique, after the eruption of Mount Pelée, August 30th 1902

The name Mount Pelée is where the word Peléan eruption comes from. Mount Pelée is in the Caribbean, on an island called Martinique. Its name is French for Bald Mountain. It is an active stratovolcano and its most famous eruption was back in 1902. This eruption is often referred to as the worst volcanic disaster of the 20th century.

The largest town on the island was called Saint Pierre and the pyroclastic flow from the 1902 eruption completely destroyed the town killing 30,000 people. The pyroclastic flow contained gases at a temperature of over 1000 C. It took only one minute for the pyroclastic flow to reach the town of Saint Pierre which was 6 km from the volcano. The temperature was so high that it set fire to everything as soon as it reached the town.

There were only two survivors in the path of the volcano. The first survivor was a man who was in jail in an underground dungeon, but he got burnt very badly and the second was a young girl who managed to escape in a fishing boat.

The town of Saint Pierre was never restored completely. A few villages were eventually rebuilt on the land.

KRAKATOA, INDONESIA

An 1888 picture of the 1883 eruption of Krakatoa

Krakatoa is a volcanic island that is between the islands of Sumatra and Java. In 1883 there was a massive eruption which also caused tsunamis. The eruption destroyed two thirds of the Krakatoa Island and was responsible for the deaths of at least 36,400 people. The explosion from the eruption has been described as the loudest sound ever heard. There are reports of the explosion being heard nearly 5000 km away.

MOUNT TAMBORA, INDONESIA

Aerial view of the caldera of Mount Tambora, formed during the colossal 1815 eruption

Mount Tambora is an active stratovolcano. In 1815 Tambora erupted, causing 160 km³ of lava, ash and other volcanic matter to come out of the volcano. This makes it the largest volcanic eruption in history. The explosion was heard 2000 km away on the island of Sumatra. At least 71,000 people were killed as a result of the eruption.

The eruption brought about a global climate change, known as a volcanic winter, because of the volcanic ash cloud which blocked the sun's rays from the Earth. Farm animals and crops could not survive in much of the northern hemisphere, including Europe and North America, during 1815 and 1816. This caused the worst famine of the 19th century and more deaths.

MOUNT VESUVIUS, ITALY

'Vesuvius in Eruption, with a View over the Islands in the Bay of Naples'

Painting by Joseph Wright of Derby, c.1776-80

As we said earlier in this book Mount Vesuvius is most famous for the eruption in AD 79 when the volcano completely destroyed the cities of Pompeii and Herculaneum. The number of people killed in this eruption is thought to be something like 18,000. Today Mount Vesuvius is known as one of the world's most dangerous volcanoes for two reasons; its eruptions tend to be Plinian eruptions which are the most deadly and also it is only 9 km from the huge city of Naples, where about three million people live. This makes the Bay of Naples in southern Italy the most densely populated volcanic region in the world.

MOUNT UNZEN, JAPAN

Mount Unzen, Japan.
In this photo you can see the flow of its lahars

Mount Unzen is actually a group of stratovolcanoes on Japan's most southerly main island. There has been volcanic activity here for thousands of years. Some of the volcanic rocks found in the area suggest to scientists that there have been volcanoes here for even longer than that, possibly from six million years ago!

The worst thing that has happened at Mount Unzen was only just over 200 years ago. In 1792 one of its lava domes collapsed and triggered a tsunami which killed about 15,000 people. This was Japan's worst ever volcanic disaster.

Mount Unzen was also active more recently. In 1991 the pyroclastic flow from an eruption killed forty-three people. Between 1991 and 1994 there were around 10,000 small pyroclastic flows which have destroyed 2000 houses. In 1995 the eruptions came to an end, but this hasn't been the end of the problems. Heavy rain in the area has caused all the debris left over from the pyroclastic flow to create lahars. In recent years dykes have been constructed to channel the flow of the lahars away from built-up areas where people live.

Devastation from the Mount Unzen 1991 eruption

NEVADO DEL RUIZ, COLUMBIA

The town of Armero was located in the centre of this photograph, taken late November 1985

The Nevado del Ruiz volcano in Columbia tends to have mostly Plinian eruptions which we know are never good news. Along with pyroclastic flows this type of eruption can also produce rivers made out of mud and all the gunge from the pyroclastic flow. These evil rivers are called lahars.

The most deadly lahar ever recorded came from just a small eruption at Nevado del Ruiz in 1985. This small eruption caused an enormous lahar that buried and destroyed the nearby town of Armero. This is now called the Armero tragedy and it caused about 25,000 deaths.

The volcano could still erupt now and scientists believe that 500,000 people in nearby towns and villages could be at risk from lahars if there is another eruption.

LAKI AND GRIMSVÖTN, ICELAND

The centre of the Laki Fissure

As we mentioned earlier in this book, Laki is a fissure vent volcano which caused global climate change during 1783, resulting in about six million deaths over the next few years. Grimsvötn is the name of a volcano next to the Laki fissure that also erupted at the same time.

SANTORINI, GREECE

Volcanic craters at Santorini today

This huge eruption happened between 1650 BC and 1500 BC. The eruption was one of the largest volcanic events in the history of the Earth. It was a Plinian eruption and is thought to have devastated the island of Santorini, as well as all the nearby islands and the coastline of Crete.

Nowadays Santorini is a glamorous holiday destination where people go to enjoy some sunshine. I wonder how many of those holidaymakers are fully aware of the devastation that happened 3500 years ago.

HUAYNAPUTINA, PERU

Ash falling on the city of Arequipa in 1600

Huaynaputina is a stratovolcano located in a volcanic upland in southern Peru. This volcano does not look like a normal volcano; it is more like a large volcanic crater. On February 19th 1600 it exploded very badly. This was the largest volcanic explosion in South America.

People watching the volcano described it as "a big explosion with cannonball-like explosions". Pyroclastic flows oozed down the mountain and some of this mixed with water from the Rio Tambo to create lahars. One hour after the eruption, ash began to fall from the sky, and within 24 hours, the city of Arequipa, 70kms away was covered with 25 cms of ash.

The eruptions kept happening until March 5th 1600 when Huaynaputina exploded. Pyroclastic flows travelled 13 km to the east and south east. Lahars destroyed several villages along the way. Ash was reported to have fallen 250,500 km away, throughout southern Peru and on what is now northern Chile and western Bolivia.

In total, the volcano killed more than 1500 people, and the ash buried ten villages. Farming was badly affected and took 150 years to get back to normal. But the worst thing about Huaynaputina was its effect on the global climate. The explosion caused temperatures to drop in the northern hemisphere where 1601 was the coldest year for 600 years, leading to The Russian Famine of

1601–1603 which is thought to have killed two million people – a third of the Russian population.

LAKE TOBA, INDONESIA

A view of Lake Toba taken on November 13th 2011, by Ken Marshall

Lake Toba is a caldera lake at the top of a supervolcano on the Indonesian island of Sumatra. It is the largest lake in Indonesia and the largest volcanic lake in the world. Lake Toba is 100 km long and 30 km wide and has a depth of 505 metres at its deepest point. Sometime between 69,000 and 77,000 years ago there was a massive eruption. When we say massive I mean really massive, in fact this is the largest eruption on Earth in the past 25 million years. When it erupted the Lake Toba volcano killed most human life on the Earth and left the population of the planet at only about 10,000 people.

It is hard to believe now that this large beautiful lake is right slap bang in the middle of a volcano that nearly wiped out the entire human race.

The next time you go on holiday somewhere, try and find out a bit more about where you are going. You can use Google and also Wikipedia to find out if you are going to be near the edge of a tectonic plate or a hotspot. If you are going to be in a volcanic zone, then see if there are any volcanoes that you can go and see.

Timanfaya National Park, Lanzarote, Canary Islands

Some popular holiday destinations are well known for their volcanic activity, such as the Canary Islands, one of my favourite spots. Lanzarote is one of the Canary Islands and the whole of the island looks like a different planet because of its last eruptions back in 1730 to 1736. Lanzarote has the Timanfaya National Park which is entirely made out of volcanic rock and soil.

At Timanfaya you can see demonstrations of the heat from a magma chamber just below the Earth's surface. The most famous is where they throw a bundle of sticks into a shallow hole and they catch fire straightaway. The 'El Diablo' restaurant serves Canarian food which is cooked using the geothermal heat from a cast-iron grill placed over a large hole in the ground!

Arthur's Seat, Edinburgh, Scotland
– did you know it was a volcano?

You don't even have to go anywhere exotic such as the Pacific Ring of Fire to find volcanoes, for example did you know that the curious mountain in Edinburgh, Scotland called Arthur's Seat was formed by an extinct volcano system from approximately 350 million years ago?

Mount Pelée, Martinique

So you see there really are volcanoes all over the world, some more interesting than others and some more dangerous than others. When I look at this photograph of holidaymakers in Martinique in the Caribbean, I wonder if they have even noticed the large mountain behind them. It is of course the active volcano Mount Pelée which erupted not so long ago in 1902. This was the worst volcanic disaster of the 20th century, killing 30,000 people.

Keep your eyes open for volcanoes, remember the Earth is not as solid as it seems!

VOLCANO DICTIONARY

There are a lot of strange (and long) words that are used when we talk about volcanoes. Throughout this book I have tried not to use too many of the more technical words, but if you research volcanoes you may come across more words you may not recognise. To make things easier you can quickly look them up here.

In the Kindle version of this book, the words that look like a hyperlink are ones you can click to see what that means too. Here it means it is also in the Volcano Dictionary, so you can easily look that word up too.

Active volcano
A volcano that has erupted sometime since the last ice age, so sometime in the last 10,000 years.

Ash
Ash is volcanic particles smaller than 2 mm in diameter that spew out of a volcano.

Ash Cloud
An ash cloud is the cloud of ash that forms after some volcanic eruptions, not to be confused with pyroclastic flow.

Basalt

Basalt is a type of volcanic (igneous) rock. This hard, dark rock is often rich in iron and magnesium. Basalt is the most common type of rock in the Earth's crust - most of the sea floor is made up of basalt.

Caldera

A caldera is a large crater formed from part of a volcano that collapses. Calderas are usually circular or elliptical. Yellowstone Caldera, Wyoming, United States is an example of a very large caldera. Caldera comes from the Spanish word for cauldron, a large pot used for cooking food and boiling liquids.

Composite volcano

A composite volcano is another name for a stratovolcano.

Dormant volcano

These are volcanoes that haven't erupted for over 10,000 years, but may erupt in the future.

Extinct volcano

An extinct volcano is a volcano that is not likely to erupt ever again.

Fissure

A fissure is a crack in a rock. A fissure vent volcano is one from which <u>lava</u> erupts.

Geology

Geology is the study of the Earth's structure, including rocks and of course, volcanoes.

Geyser

A geyser looks like a sudden jet of water and steam coming out of the Earth. Geysers are mostly found near active volcanic areas, and they happen because a <u>magma chamber</u> is nearby. Water from the surface of the Earth makes its way down cracks or holes to a depth of around 2,000 metres where it comes into contact with hot rocks. The pressurized water boils instantly and steam is seen spraying out of the Earth.

Hotspot

A hotspot is a thin area in a <u>tectonic plate</u> where magma rises. Volcanoes often erupt over hotspots, such as the Hawaiian volcanoes.

Igneous rock

When molten volcanic rock cools, the rock that is formed is called <u>igneous</u>. Some igneous rocks include granite, <u>volcanic glass</u> and <u>basalt</u>.

Java (Indonesia)

Java is almost entirely made of volcanic rock, there are lots of volcanoes there and forty-five of them are considered active volcanoes. As is the case for many other Indonesian islands, volcanoes have played an important part in the geological and human history of Java.

Krakatoa

Krakatoa is a volcano located in Indonesia. On August 26th 1883 Krakatoa erupted, killing thousands of people. This was one of the biggest volcanic eruptions in modern times.

Lahar

A lahar is a moving mixture of rock, water, and other debris that falls down the slopes of a volcano and/or a river valley. Lahar is an Indonesian word.

Lapilli

Lapilli is a name for volcanic cinders; this is like the ash from a volcano, but in larger particles from 2mm diameter to 64 mm in diameter.

Lava

Lava is magma that has left the volcano. Lava will eventually cool down and form igneous rock.

Magma

Magma is molten rock from which igneous rock forms. Magma is formed from many types of rocks, including basalt. Once the magma has left the volcano it is then known as lava.

Magma chamber

A magma chamber is like a big underground pool of molten rock at the base of a volcano ready to spurt out of the top. When a lot of the magma has gone from the magma chamber, the top of the volcano collapses and forms a caldera.

Mercury

As well as being found in thermometers, mercury exists mostly in the Earth's crust. Mercury is a metal that is a liquid at room temperature. It is formed when cinnabar rock (mercury ore) is heated, so guess what? This happens a lot in volcanoes.

Neck

This is the solidified lava that fills the top of a volcano and eventually causes the volcano to stop erupting. A neck is also called a volcanic plug.

(volcano) Observatory

A volcano observatory is where scientists research and monitor volcanoes. They continuously look at changes in atmospheric conditions and volcanic activity between and during eruptions. While an eruption is happening, the observatory will supply up to date information about the volcano.

Pyroclastic flow
Pyroclastic flow is an avalanche of pyroclastic material, broken rock, pumice, ash and hot gases that erupt from a volcano. A pyroclastic flow travels fast and is very hot.

Pumice
A volcanic rock made of rough textured volcanic glass, which may or may not contain crystals. It is typically pale in colour, ranging from white, cream, blue or grey, to green-brown or black. It is used a lot in building work and you can even use a 'pumice stone' to get rid of the hard skin from the bottom of your feet!

Quicksilver
Another name for the liquid metal, mercury found in the Earth's crust. This metal can be formed by the heat of a volcano.

Quartz
Quartz crystals can be sometimes found in igneous rock, this is an example of xenocryst.

Quick, run!
As daft as it may sound, if you are being chased by lava flow this is good advice as it usually moves at walking

pace. If you are running away from a lahar you might stand a chance if you can get out of the path of the lahar. With pyroclastic flow you have no chance, run if you like, but it won't help.

Ring of Fire

The Ring of Fire is an area around the Pacific Ocean that has lots of volcanic activity and lots of volcanoes.

Shield volcano

A shield volcano is a volcano that has gently sloping sides. Shield volcanoes are made mostly of basalt.

Stratovolcano

Stratovolcanoes are tall and conical and are made in layers. Stomboli in Italy is a good example of a stratovolcano. Another name for a stratovolcano is a composite volcano.

Supervolcano

A supervolcano is an enormous volcano much larger than any other type of volcano. A supervolcano forms when a huge magma chamber in the Earth's crust erupts after being under great pressure, causing a large caldera to form as the land over the magma chamber collapses. The biggest volcano in the world is the Yellowstone Supervolcano in the United States.

Tectonic plates

Tectonic plates are huge chunks of the Earth's crust that form the surface of the Earth. Volcanoes often form where tectonic plates meet, but can also form in the middle of a tectonic plate at a hotspot.

Tephra

Tephra is the volcanic material that is made of ash, pyroclastic flow and other materials when they reach the ground but remain as volcanic fragments and don't turn into igneous rock. Tephra fragments are classified by size as ash, lapilli and volcanic bombs.

United States

The United States of America has some amazing volcanoes and includes the world's largest volcano, the Yellowstone Supervolcano. The west coast of the United States and Alaska is part of the Ring of Fire.

Volcanic Bombs

Volcanic bombs are any particles larger than 64mm that are thrown out of an erupting volcano.

Volcanic Glass

Volcanic glass forms when magma cools very quickly. Like all types of glass, it is not crystalline. Volcanic glass can refer to several different kinds of igneous rock.

Volcanologist

A volcanologist is a scientist who studies volcanoes.

Whangaehu River

This is where the terrible Tangiwai disaster happened caused by a lahar. This was New Zealand's worst ever rail accident. The Whangaehu River Bridge collapsed beneath an express passenger train at Tangiwai, in the central North Island of New Zealand. The locomotive and first six carriages derailed into the river, killing 151 people.

Xenocryst

A Xenocryst is a crystal that forms inside igneous rock, such as quartz crystal

Yellowstone

The Yellowstone Supervolcano is the biggest volcano in the World. The Yellowstone Caldera attracts thousands of visitors every year as it part of the Yellowstone National Park in Wyoming, United States. It is thought that it may erupt again one day, but not for a very long time. If it does explode it could be a thousand times more powerful than the Mount St Helens eruption in 1980.

(volcano) Zones

Volcano zones are where volcanoes form; often these are places where tectonic plates meet or over hotspots.

The Volcano Book for Children, Mums, Dads and Teachers

All the images in this book are listed here, with a little bit more information about them. As much as I would have loved to have spent a few years of my life travelling around the world taking photos of volcanoes, this just wasn't possible. Luckily there are plenty of people who have done this and have been good enough to allow anyone to share their photographs as long as you say who they are, so these are listed here. These are usually called 'Creative Commons' images. As well as this there are a lot of old images that are just available for anyone to use as they like, they are listed here too. These are called 'Public Domain' images. You might find this list useful if you are doing your own project about volcanoes, but otherwise if you look up some of the photographers on the internet you will find lots of other images that you can use too.

Front Cover
This is the Puʻu ʻŌʻō cinder cone, part of the Kilauea volcano of the Hawaiian Islands. It was taken by Brian Snelson. Brian said 'We were flying away to the coast when the pilot said she could see that the volcano was erupting so she turned back. As she banked for me to take photos we could feel the heat." The photo was taken in October 1997.
This work is licensed under a Creative Commons Attribution 2.0 Generic License

Rick and Madeline Lomas on holiday in Lanzarote (Volcano Island)
30th October 2005
The author's photograph. Rick and Madeline Lomas on the Romero ferry at Órzola, Lanzarote, about to sail to the island of La Graciosa. Photo taken by Susan Lomas.

The Earth has layers, quite a few actually!
Earth and atmosphere cutaway illustration by Washiucho.

The magma chamber is like a big underground pool of molten rock
Blank scheme of a vulcanian eruption. By Sémhur [FAL or CC-BY-SA-3.0-2.5-2.0-1.0 (http://creativecommons.org/licenses/by-sa/3.0)], via Wikimedia Commons.

The Earth's crust is made of large pieces called tectonic plates
This is a NASA image modified by Rick Lomas. The original image This file is in

The Volcano Book for Children, Mums, Dads and Teachers

the public domain because it was solely created by NASA. NASA copyright policy states that "NASA material is not protected by copyright unless noted.

This map shows 15 of the largest tectonic plates, where do you live?
This map shows 15 of the largest plates. Note that the Indo-Australian Plate may be breaking apart into the Indian and Australian plates, which are shown separately on this map. February 1996 by USGS.

The 1982 eruption of Galungung in western Java, Indonesia
By R. Hadian, U.S. Geological Survey (image from NOAA website) [Public domain], via Wikimedia Commons

La Cumbre volcano, Fernandina Island, Galapagos. A perfect caldera
Photographed from the International Space Station. This photo was created by the Image Science & Analysis Laboratory, of the NASA Johnson Space Center.

Crater Lake is a caldera lake in Oregan, USA.
Crater Lake is a caldera lake in the U.S. state of Oregon. It is the main feature of Crater Lake National Park and famous for its deep blue color and water clarity. The lake partly fills a 1,220 mdeep caldera that was formed by the collapse of the volcano Mount Mazama. Taken by Zainubrazvi on Monday, July 17, 2006.

The Pacific Ring of Fire
A public domain image by Gringer, 11 February 2009

An active volcano, Mount Etna erupting in 2011
Taken on July 30, 2011 by gnuckx on Flickr

A dormant volcano, Putauaki in New Zealand
Taken on 14 December 2006 by Kahuroa
Released into public domain, no rights reserved

Ten metre high fountain of lava, Hawaii, United States
Arching fountain of lava approximately 10 m high. South of Pu'u Kahaualea. By J.D. Griggs, 10th September 2007. This image is in the public domain because it contains materials that originally came from the United States Geological Survey, an agency of the United States Department of the Interior.

Pyroclastic flows at Mayon Volcano, Philippines, 1984
Pyroclastic flows descend the south-eastern flank of Mayon Volcano,

Philippines. There were no casualties from the 1984 eruption because more than 73,000 people evacuated the danger zones as recommended by scientists of the Philippine Institute of Volcanology and Seismology. A public domain photo from 23 September 1984 by C.G. Newhall.

Vesuvius as seen from the ruins of Pompeii, which was destroyed in the eruption of AD 79
This file is licensed under the Creative Commons Attribution-Share Alike 3.0 Unported license, by Morn the Gorn, 15th March 1998.

A lahar from the 1982 eruption of Galunggung.West Java, Indonesia
These lahars from the 1982 eruption of Galunggung volcano on the Indonesian island of Java caused extensive damage to houses and croplands.
Taken in 1983 by Robin Holcomb, U.S. Geological Survey.
This image is in the public domain because it contains materials that originally came from the United States Geological Survey, an agency of the United States Department of the Interior.

Ash plume from Mt Cleveland, Alaska
Eruption of Cleveland Volcano, Aleutian Islands, Alaska is featured in this image photographed by an Expedition 13 crewmember on the International Space Station. This file is in the public domain because it was created by the Image Science & Analysis Laboratory, of the NASA Johnson Space Center. 23rd May 2006.

The Laki Fissure in Iceland, not the prettiest of volcanos, but one of the most deadly!
Sight to the central fissure of Laki volcano, Iceland. This file is licensed under the Creative Commons Attribution-Share Alike 3.0 Unported license. By Chmee2/Valtameri, 3rd August 2009.

The volcano Skjaldbreiður, Iceland, a perfect example of a shield volcano
This file is licensed under the Creative Commons Attribution-Share Alike 3.0 Unported license. By Reykholt, October 2004.

Lava domes in the crater of Mount St. Helens
Mount St. Helens and Crater Glacier, Cascade Range, Washington, United States. Public domain photo by Willie Scott, USGS, 12 September 2006.

The bulging cryptodome on the side of Mt. St. Helens on 27th April 1980
A "bulge" developed on the north side of Mount St. Helens as magma pushed up within the peak. Angle and slope-distance measurements to the bulge indicated it was growing at a rate of up to 1.5 meters per day. By May 17th, part of the volcano's north side had been pushed upwards and outwards over 135 meters. The view is from the northeast.

Sunset Crater near Flagstaff, Arizona, USA
Sunset Crater National Monument, Arizona, USA. Cinder cone. This image contains material based on a work of a National Park Service employee, created as part of that person's official duties. As a work of the U.S. federal government, such work is in the public domain.

Stromboli, Italy is good example of a stratovolcano
Photo taken by Patrick Nouhailler on 20th June 2013
This file is licensed under the Creative Commons Attribution-Share Alike 3.0 Unported license.

The Yellowstone Caldera is the volcanic caldera and supervolcano located in Yellowstone National Park in the United States, sometimes referred to as the Yellowstone Supervolcano
Yellowstone River in Hayden Valley. Yellowstone National Park, Wyoming, USA. Photo by Ed Austin/Herb Jones 1987. This image or media file contains material based on a work of a National Park Service employee, created as part of that person's official duties. As a work of the U.S. federal government, such work is in the public domain.

The Sonoma geothermal power plant at The Geysers field in the Mayacamas Mountains, Somona County California, USA
The Sonoma Calpine 3 geothermal power plant at The Geysers field in the Mayacamas Mountains of Somona County California, USA. Photographed looking northwest from the nearby helipad. 14 April 2012 by Stepheng3.

Stomboli
Photo taken by Patrick Nouhailler on 20th June 2013
This file is licensed under the Creative Commons Attribution-Share Alike 3.0 Unported license.
Patrick has some very interesting photographs on Flickr.com, please do take a look

Red hot lava oozing into the ocean at Kilauea Volcano, Hawaii
a photo taken by Bob Webster on 4th April 2008
This photo is from Moa Heiau, Hawaii, US. Please check out Bob's other photos
on Flickr.com

Burnt out trees in a lava field at Etna, taken by 'gnuckx' on 1 May 2009
The Flickr contributor 'gnuckx' has some other really good Etna photographs
that are well worth a look at.

Space radar image of Teide, taken in 1994 by NASA
This radar image shows the Teide volcano on the island of Tenerife in the
Canary Islands. The image was acquired by the Spaceborne Imaging Radar
onboard the space shuttle Endeavour on October 11, 1994. NASA copyright
policy states that "NASA material is not protected by copyright unless noted.

**Mount Ruapehu and the Chateau Tongariro Taken on September 14th 2009
by Sid Mosdell**
Sid Mosdell has some wonderful photographs on Flickr.com which are well
worth looking at.

Mount Fuji, Japan, photo taken on 3 May 2010 by Ivan Walsh
At the time of writing this book Ivan Walsh has a collection of 161 photographs
of Mount Fuji on his Flickr.com account and they are all pretty amazing.

Cotopaxi Volcano, photo taken on 26th September 2009 by Ivar Abrahamsen
Ivar Abrahamsen has the username of 'flurdy' on Flickr.com and his photos are
well worth checking out.

**The Sierra Negra Volcano, a photo taken by Michael R Perry on the 23rd
December 2009**
On Flickr.com Michael R Perry says, "The slopes of Sierra Negra Volcano are
first misty, then crystal clear as we hike across into lava flows from eruptions in
1970s, 80s and most recently, 2005.

**A geyser – where water is turned instantly to steam, at Yellowstone, 13th
October 2000**
A photo from Flickr.com by the contributor InSapphoWeTrust. Who says, "…
lots of geysers are found throughout the landscape. This is one. Normally they
will billow steam like this, but occasionally they may erupt.

Buildings destroyed by the volcano near Plymouth, Montserrat. 30th April 2012

Photo by Christine Warner Hawks at Old Towne, Saint Peter, Montserrat. Christine has a photo set on Flickr.com called 'Antigua and Montserrat Adventure' from her trip in 2012. There are some quite incredible photographs of the volcanic destruction.

Main Street, Morne Rouge, Martinique, after the eruption of Mount Pelée, Aug 30, 1902

This is special kind of print called a lithograph from the Library of Congress Prints and Photographs. This media file is in the public domain in the United States. This applies to U.S. works where the copyright has expired, often because its first publication occurred prior to January 1, 1923.

An 1888 picture of the 1883 eruption of Krakatoa

This image is in the public domain because its copyright has expired.

Aerial view of the caldera of Mount Tambora, formed during the colossal 1815 eruption

A photograph taken by Jialiang Gao (peace-on-earth.org) from the air on the 7th June 2011 at the island of Sumbawa, Indonesia

Vesuvius eruption, a painting by Joseph Wright of Derby

This is a painting from the Art collection of the Huntington Library in Pasadena. By Joseph Wright of Derby (1734–1797)
This work is in the public domain.

Mount Unzen, here you can see the flow of lahars

This work is in the public domain in the United States because it is a work prepared by an officer or employee of the United States Government as part of that person's official duties

Devastation from the Mount Unzen 1991 eruption

Devastation resulting from eruption of Mt. Unzen, Nagasaki Pref., Japan
User Fg2 took this photograph and contributed it to the public domain.

The town of Armero was located in the center of this photograph, taken late November 1985

Río Lagunillas, former location of Armero. An eruption from Ruiz's summit on

November 13, 1985, at 9:08 p.m. sent a series of pyroclastic flows and surges across the volcano's ice-covered summit. This caused lahars to flow. A photograph taken by Jeffrey Marso, a USGS geologist. This image is in the public domain because it contains materials that originally came from the United States Geological Survey, an agency of the United States Department of the Interior.

Center of the Laki Fissure
The central fissure of Laki volcano, Iceland. Taken on the 3rd August 2009 by Chmee2/Valtameri
This file is licensed under the Creative Commons Attribution-Share Alike 3.0 Unported license.

Volcanic craters at Santorini today
Nea Kameni, Greece. Taken on the 5th June 2011 by Tango7174
Permission is granted to copy, distribute and/or modify this document under the terms of the GNU Free Documentation License, Version 1.2.

Ash falling on the city of Arequipa in 1600
The City of Arequipa during the catastrophic eruption of the volcano Huaynaputina (1600). Tons of ashes fall upon the city. A drawing by Guaman Poma.
This image is in the public domain because its copyright has expired.

A view of Lake Toba taken on November 13, 2011 by Ken Marshall
Taken on November 13, 2011 from Onan Runggu Timur, North Sumatra, ID

Arthur's Seat, Edinburgh, Scotland – did you know it was a volcano?
Taken by Kim Traynor on 15th February 2012
This image is licensed under the Creative Commons Attribution-Share Alike 3.0 Unported license.

Timanfaya National Park, Lanzarote, Canary Islands
Taken by Gernot Keller, London (www.gernot-keller.com) on 19 December 2008.This image is licensed under the Creative Commons Attribution-Share Alike 2.5 Generic license.

A final word

If you have enjoyed reading this book please consider leaving a review on Amazon. Reviews help other people to decide if they should buy the book or not as well as proving valuable feedback for the author.

RickLomas.com

43993270R00058

Made in the USA
Middletown, DE
25 May 2017